U.S. Department
of Transportation

Federal Highway
Administration

FHWA TRAFFIC NOISE MODEL®
USER'S GUIDE
(VERSION 2.0 ADDENDUM)

Final Report
March 2002

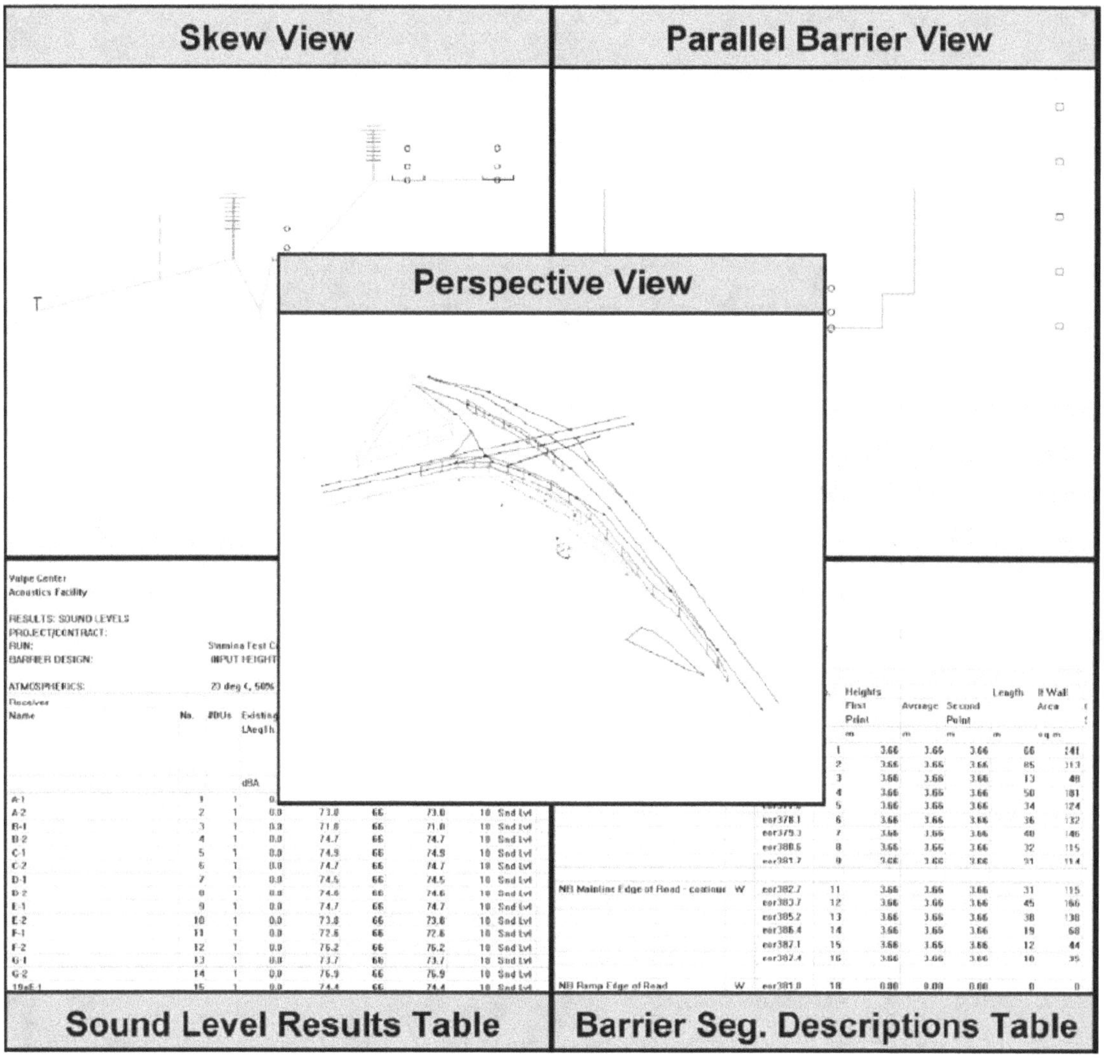

Prepared for

U.S. Department of Transportation
Federal Highway Administration
Office of Natural Environment
Washington, DC 20590

Prepared by

U.S. Department of Transportation
Research and Special Programs Administration
John A. Volpe National Transportation Systems Center
Environmental Measurement and Modeling Division, DTS-34
Cambridge, MA 02142-1093

NOTICE

This document is disseminated under the sponsorship of the Department
of Transportation in the interest of information exchange. The United
States Government assumes no liability for its contents or use
thereof. This report does not constitute a standard, specification,
or regulation.

The United States Government does not endorse products or
manufacturers. Trade or manufacturers' names appear herein solely
because they are considered essential to the object of this document.

REPORT DOCUMENTATION PAGE

Form Approved
OMB No. 0704-0188

Public reporting burden for this collection of information is estimated to average 1 hour per response, including the time for reviewing instructions, searching existing data sources, gathering and maintaining the data needed, and completing and reviewing the collection of information. Send comments regarding this burden estimate or any other aspect of this collection of information, including suggestions for reducing this burden, to Washington Headquarters Services, Directorate for Information Operations and Reports, 1215 Jefferson Davis Highway, Suite 1204, Arlington, VA 22202-4302, and to the Office of Management and Budget, Paperwork Reduction Project (0704-0188), Washington, DC 20503.

1. AGENCY USE ONLY (Leave blank)	2. REPORT DATE March 2002	3. REPORT TYPE AND DATES COVERED Final Report March 1998 - March 2002
4. TITLE AND SUBTITLE FHWA Traffic Noise Model® User's Guide (Version 2.0 Addendum)		5. FUNDING NUMBERS HW266/H2008
6. AUTHOR(S) Cynthia S.Y. Lee, Judith L. Rochat, and Gregg G. Fleming		
7. PERFORMING ORGANIZATION NAME(S) AND ADDRESS(ES) U.S. Department of Transportation Research and Special Programs Administration John A. Volpe National Transportation Systems Center Environmental Measurement and Modeling Division, DTS-34 Cambridge, MA 02142-1093		8. PERFORMING ORGANIZATION REPORT NUMBER
9. SPONSORING/MONITORING AGENCY NAME(S) AND ADDRESS(ES) U.S. Department of Transportation Federal Highway Administration Office of Natural Environment Washington, DC 20590		10. SPONSORING/MONITORING AGENCY REPORT NUMBER

11. SUPPLEMENTARY NOTES
FHWA Program Managers: Robert E. Armstrong and Steven A. Ronning, HENE, Office of Natural Environment

12a. DISTRIBUTION/AVAILABILITY STATEMENT	12b. DISTRIBUTION CODE

13. ABSTRACT (Maximum 200 words)

In March 1998, the Federal Highway Administration (FHWA) Office of Natural Environment, released the FHWA Traffic Noise Model (FHWA TNM) Version 1.0, a state-of-the-art computer program for highway traffic noise prediction and analysis. Since then, the FHWA, with assistance from the Volpe Center Acoustics Facility (Volpe Center) and Foliage Software Systems (FSS), have released updates of TNM (Versions 1.0a, and 1.0b) in March 1999 and August 1999, respectively. In support of the FHWA and the California Department of Transportation, the Volpe Center and FSS released Version 1.1 in September 2000. TNM 2.0 is the latest release of the TNM software. Two companion reports were released with TNM Version 1.0, a Technical Manual that describes the acoustics within TNM and a User's Guide. In addition, prior to TNM release, a data report was published that describes the vehicle noise-emissions data base within TNM. This document is an addendum to the FHWA TNM Version 1.0 User's Guide. It details the enhancements in the program up to and including Version 2.0.

14. SUBJECT TERMS Highway traffic-noise prediction, FHWA TNM, digitizer, roadways, traffic, receivers, barriers, building rows, terrain lines, ground zones, tree zones, noise barriers			15. NUMBER OF PAGES 35
			16. PRICE CODE
17. SECURITY CLASSIFICATION OF REPORT Unclassified	18. SECURITY CLASSIFICATION OF THIS PAGE Unclassified	19. SECURITY CLASSIFICATION OF ABSTRACT Unclassified	20. LIMITATION OF ABSTRACT

NSN 7540-01-280-5500

Standard Form 298 (Rev. 2-89)
Prescribed by ANSI Std. 239-18
298-102

PREFACE

In March 1998, the Federal Highway Administration (FHWA) Office of Natural Environment, released the FHWA Traffic Noise Model (FHWA TNM) Version 1.0, a state-of-the-art computer program for highway traffic noise prediction and analysis. Since then, the FHWA, with assistance from the Volpe Center Acoustics Facility (Volpe Center) and Foliage Software Systems (FSS), have released updates of TNM (Versions 1.0a, and 1.0b) in March 1999 and August 1999, respectively. In support of the FHWA and the California Department of Transportation, the Volpe Center and FSS released Version 1.1 in September 2000. TNM 2.0 is the latest release of the TNM software. Two companion reports were released with TNM Version 1.0, a Technical Manual that describes the acoustics within TNM and a User's Guide.[1,2] In addition, prior to TNM release, a data report was published that describes the vehicle noise-emissions data base within TNM.[3]

This document is an addendum to the FHWA TNM Version 1.0 User's Guide. It details the enhancements in the program up to and including Version 2.0.

TABLE OF CONTENTS

Section   **Page**

1. **GETTING STARTED** 1
   - 1.1 How to Use This User's Guide Addendum 1
   - 1.2 Contents of the TNM Version 1.1 Package 1
   - 1.3 Hardware and Software requirements 2
   - 1.4 Installation 3
   - 1.5 Source Code Licensing Agreement 4
   - 1.6 Technical Support 4

2. **WHAT'S NEW** 5
   - 2.1 File Menu 5
     - 2.1.1 New 5
     - 2.1.2 Open 5
     - 2.1.3 Import a STAMINA 2.0/OPTIMA File 6
     - 2.1.4 Import a DXF File 6
     - 2.1.5 Cleanup Run 8
   - 2.2 Edit Menu 9
     - 2.2.1 Move 9
   - 2.3 View Menu 10
     - 2.3.1 Show/Hide 10
   - 2.4 Setup Menu 10
     - 2.4.1 General 10
   - 2.5 Input Menu 11
     - 2.5.1 Roadway Input 11
     - 2.5.2 Receiver Input 11
     - 2.5.3 Contour Zone 14
     - 2.5.4 Adjustment Factors 14
     - 2.5.5 Input Check 14
     - 2.5.6 User-Defined Vehicles 14
   - 2.6 Calculate Menu 15
     - 2.6.1 Error-Catching Mechanism 15
     - 2.6.2 Multiple Runs 16
   - 2.7 Contours Menu 19
     - 2.7.1 Calculating Contours 19

2.8 Tables Menu 19
    2.8.1 All Results Tables 19
    2.8.2 Print Tables 20
    2.8.3 Barrier Design Table 21

3. CERTIFIED OUTPUT FOR THE OFFICIAL TNM TEST CASE 23

REFERENCES 25

LIST OF FIGURES

Figure   **Page**

Figure 1. TNM run-time comparison. 3
Figure 2. Convert Run dialog. 5
Figure 3. Cleanup Run menu item. 9
Figure 4. Cleanup Run dialog. 9
Figure 5. Cleanup Run "select run" dialog. 9
Figure 6. Changes to Receiver Input dialog: General tab. 11
Figure 7. Changes to Receiver Input dialog: Levels/Criteria tab. 12
Figure 8. Changes to Receiver Input dialog: Adjustment Factors tab. 12
Figure 9. Changes to Receiver Input dialog: Notes tab. 12
Figure 10. Changes to receiver input table. 13
Figure 11. Changes to sound level results table. 13
Figure 12. Invalidated receivers due to floating point errors. 15
Figure 13. Calculate menu. 16
Figure 14. Calculation Manager dialog. 17
Figure 15. Browse for Folder dialog. 17
Figure 16. Continuing calculations after cancelling. 17
Figure 17. Multiple run (batch-mode) capability: output report. 18
Figure 18. Version identification in all results tables. 20
Figure 19. Print Tables menu item. 20
Figure 20. Print Tables dialog. 20
Figure 21. Barrier Design Table menu item. 21
Figure 22. Barrier design table: expanded display. 22
Figure 23. Barrier design table: condensed display. 22

Figure 24. Updated sound level results table for the official TNM test case. 23

LIST OF TABLES

Table   **Page**

Table 1. DXF Import Items. 6

1. GETTING STARTED

This section lists TNM's hardware and software requirements and provides instructions on how to install TNM.

1.1 How to Use This User's Guide Addendum

This User's Guide Addendum is essential to both the experienced and inexperienced TNM user. It details the enhancements in the program up to and including Version 2.0. Use it in addition to the TNM 1.0 User's Guide, your main information source. New users should also use the "TNM Trainer," the interactive tutorial that is included in the TNM Version 1.0 package.[4] For quick reference help, users can select the Help menu while using TNM. In addition, TNM sometimes "pops up" brief help information during various operations (see Section 7.1 in the TNM 1.0 User's Guide).

The different typeface and icon conventions used in this User's Guide Addendum are as follows:

Bold Bold is used for emphasis and to introduce new terms, which are defined in Section 2, Terminology, of the TNM 1.0 User's Guide.

Underline Underline is used to denote an available shortcut key to invoke a menu or submenu item. For example, to invoke the Help menu while using TNM, press the Alt+H keys.

A light bulb icon points out helpful tips, suggestions, engineering hints, and shortcuts that may save you time.

A stop sign icon represents warnings and when you should pay special attention to what you're doing to avoid unexpected results.

A Department of Transportation icon points out when you should refer to Appendix A of the TNM 1.0 User's Guide for FHWA policies relating to a topic.

1.2 Contents of the TNM Version 1.1 Package

The TNM Version 2.0 package contains the following:
- The FHWA TNM Version 2.0 software; and
- This FHWA TNM User's Guide Addendum.

Note: TNM Version 2.0 is the latest upgrade to the TNM Version 1.0 release package. It is a complete release package. Existing owners of TNM may purchase the latest upgrade, TNM 2.0, at a reduced cost.

Information on how to purchase the TNM can be found on the McTrans website (http://www-mctrans.ce.ufl.edu) or by contacting McTrans at:

    McTrans Center   Telephone: (352) 392-0378
    University of Florida   Fax: (352) 392-3224
    2088 Northeast Waldo Road
    Gainesville, FL 32609

Additional information on TNM 2.0 can be found on the official TNM website (http://www.tiac.net/users/a1f04/tnm). The webpage contains this TNM Version 2.0 User's Guide Addendum in Adobe Acrobat PDF file format, as well as contact information for technical support.

1.3 Hardware and Software requirements

Beginning with the release of Version 1.1, TNM has been upgraded from a 16-bit Microsoft® Windows application to a 32-bit Windows application. This change effectively removes any platform dependence, allowing TNM to run more efficiently regardless of the Windows operating system (Windows 95, 98, NT, etc.). It should be noted that this change has also obviated TNM use on previous Windows operating systems (e.g., Windows 3.1, since 3.1 is a 16-bit platform).

Substantial improvements to computational run-time have been shown during testing with most cases computing in approximately half the time. A run-time comparison of the official TNM test case (see Section 3) is shown below for several combinations of hardware and operating systems. The test case consists of the following: 9 roadways, 1 barrier (with 3 perturbations up and down), 2 terrain lines, and 32 receivers.

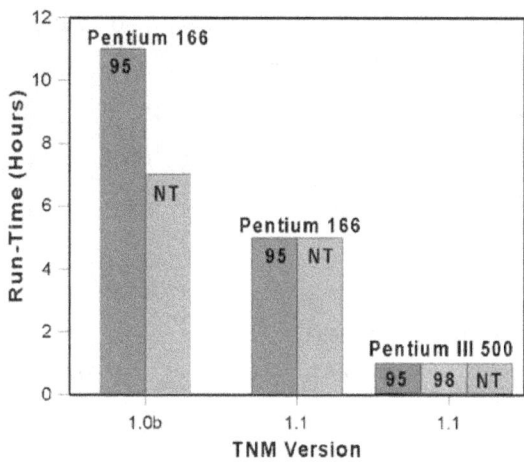

Figure 1. TNM run-time comparison.

The recommended computer system requirements for TNM Version 2.0 are:

- Computer: IBM-compatible PC;
- Processor: 500 MHZ Pentium (or faster);
- Memory: 32 MB (or more);
- Disk Drive: CD-ROM;
- Mouse input device;
- Monitor: Accelerated Super VGA (1024 x 768), 16 colors, configured with "small" fonts;
- Software: Microsoft® Windows 95 (or later) or Windows NT 4 (or later);
- 10 MB of hard-disk space for the TNM system (including sample runs); and
- Up to 1 MB of hard-disk space for each TNM run.

Although TNM may run on a less efficient computer than recommended above, some capabilities may be affected, including the graphical user interface or the speed of the noise calculations. For additional information, refer to Section 1.3 in the TNM 1.0 User's Guide.

1.4 Installation

To install TNM on your computer:
1. Insert the TNM Version 2.0 CD-ROM into your CD-ROM drive.
2. Run InstlTNM.exe.

3. When the Custom Installation box appears on your screen, select Set Location to tell the setup program where to locate TNM on your hard drive. Note that TNM 2.0 installation will not interfere with previous version of TNM installed on your computer.
4. Then after the setup program is done, configure your computer display to work with TNM. For additional information, refer to Section 1.4 in the TNM 1.0 User's Guide.

1.5 Source Code Licensing Agreement

The FHWA TNM is a registered copyright and trademark, which encompasses the User's Guide, Technical Manual, and software source and executable codes. For developers interested in obtaining the software source code, acceptance of the TNM Source Code Licensing Agreement is a prerequisite. Under the terms of the Licensing Agreement, any modifications, enhancements, or derivatives of TNM, as well as distribution of the modified source code, which makes reference to the FHWA's trademarks, are strictly prohibited without the express written permission of the FHWA.

 See Appendix A in the TNM 1.0 User's Guide for FHWA policy related to TNM Copyright and Trademark.

1.6 Technical Support

Services are available to help you with your questions. Registered owners are entitled to receive technical support and information on upgrades and supplementary guides. For installation and supplementary guide information, users may contact the Center for Microcomputers in Transportation (McTrans) at (352) 392-0378 or on the World Wide Web (http://www-mctrans.ce.ufl.edu). For technical support, a Frequently Asked Questions website is available (http://www.tiac.net/users/a1f04/tnm/faq.htm). Users may also contact the Volpe Center, Acoustics Facility at (617) 494-2372, or FHWA at (202) 366-2073.

2. WHAT'S NEW

This section discusses the enhancements that have occurred in TNM since Version 1.0. It includes changes implemented in Versions 1.0a, 1.0b, 1.1, and 2.0. The structure of this entire section follows the order of menu items within TNM:

- File Menu: Section 2.1;
- Edit Menu: Section 2.2;
- Setup Menu: Section 2.3;
- View Menu: Section 2.4;
- Input Menu: Section 2.5;
- Calculate Menu: Section 2.6;
- Contours Menu: Section 2.7; and
- Tables Menu: Section 2.8.

2.1 File Menu

2.1.1 New. When creating a new run in TNM Version 2.0, it is important to note that because TNM has been upgraded from a 16-bit Windows application to a 32-bit Windows application since TNM 1.1, the eight-character run-name restriction no longer exists. Extended run file names may be used for Windows 95 (or later) and NT Version 4.0 (or later).

 New TNM Version 2.0 runs: It is also important to note that any run created using TNM Version 2.0 will not run in 16-bit versions of TNM (e.g., Versions 1.0, 1.0a, and 1.0b).

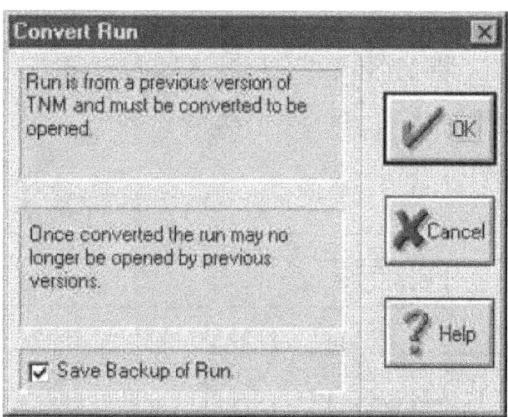

2.1.2 Open. Opening a Version 1.0, 1.0a, or 1.0b TNM run will display the **Convert Run** dialog is displayed (see Figure 2). Once the user selects OK, TNM will automatically convert and open the selected run into a Plan View. Note that if you ever want to use the original unconverted run, you must save a backup of the run before selecting the OK button in the Convert Run dialog. The "Save Backup of Run"

Figure 2. Convert Run dialog.

checkbox in the dialog allows you to automatically save a copy of your run as "Backup of Run_Name" prior to TNM Version 2.0 conversion.

 Save a backup of your runs: Always save a backup copy of your original TNM run prior opening it in TNM 2.0. If you do not save a backup copy of the run, then once that run has been converted by TNM 2.0, it cannot be opened in previous versions of TNM.

2.1.3 Import a STAMINA 2.0/OPTIMA File. TNM allows you to import a STAMINA 2.0/OPTIMA file. Previous versions of TNM did not correctly import STAMINA shielding factors. This was been corrected since TNM Version 1.1. To import shielding factors into TNM, be sure to select the **Import Shielding Factors** check box in the **Import STAMINA 2.0 Input File** dialog window (see Section 4.7.1 in the TNM 1.0 User's Guide for more details).

 Shielding factors and adjustment factors: TNM imports shielding factors from your STAMINA files as adjustment factors (see Section 8.4.4 in the TNM 1.0 User's Guide). Typically, shielding factors in STAMINA files were used to account for things such as building row and tree zone effects. However, you may choose not to import them, but model building rows and tree zones explicitly in TNM. You may wish to input/edit TNM adjustment factors for model calibration to account for parallel barrier degradations between receiver-roadway pairs and propagation effects not calculated by TNM – for example, wind effects.

2.1.4 Import a DXF File. TNM's DXF import functionality was substantially updated in TNM 1.1 to support compatibility with AutoCAD® 2000 objects. The following table shows all the DXF objects available in AutoCAD® 2000 and indicates which items TNM Version 2.0 can and cannot import. Note that shaded items in the table are items that previous versions of TNM could not import, but have been available since Version 1.1.

Table 1. DXF Import Items.

DXF Item	Import Object for Conversion	Import as Background
3DFace	No	Yes
3DSolid	No	No
ACAD_Proxy_Entity	No	No

Table 1. DXF Import Items.

DXF Item	Import Object for Conversion	Import as Background
Arc *	Yes	Yes
Arcaligned text **	No	Yes
Attdef	No	No
Attrib	No	No
Body	No	No
Circle *	Yes	Yes
Dimension **	Yes	Yes
Ellipse	Yes	Yes
Hatch	No	No
Image	No	No
Insert **	Yes	Yes
Leader	No	No
Line	Yes	Yes
LWPolyline	Yes	Yes
Mline	No	Yes
Mtext **	No	Yes
Oleframe	No	No
Ole2frame	No	No
Point	No	No
Polyline ***	Yes	Yes
Ray	No	No
Region	No	No
Rtext	No	No
Seqend	Yes	Yes
Shape	No	No
Solid	No	Yes
Spline	Yes	Yes
Text	No	No
Tolerance	No	No
Trace	No	Yes
Vertex	No	No
Viewport	No	No

Table 1. DXF Import Items.

DXF Item	Import Object for Conversion	Import as Background
Wipeout	No	No
Xline	No	No

* Arcs and circles are converted into a series of straight segments.

** When you import a DXF file for conversion into TNM, DXF labels and other text are placed in the DXF Background, which is not normally displayed by default. To view labels/text, you must select Show/Hide in the View menu, then check the "DXF Background" check box in the "Show Objects" column (see also Section 6.2 in the TNM 1.0 User's Guide).

*** Polylines are imported if the polyline or any of its vertices are complex, i.e., curve-fit, spline-fit, mesh, or polyface mesh.

Importing DXF Point Objects: Currently, TNM does not import DXF points. As a work-around, the user may connect points with polylines in the CAD program prior to import, then after import, snap-digitize TNM receivers to the DXF points in the polylines.

Importing metric DXF files: In TNM prior to Version 2.0, TNM had been importing metric DXF files incorrectly scaled by a factor of 3.281 (the scaling factor for converting metric to feet) regardless of the default TNM units setting (see Section 7.2 in the TNM 1.0 User's Guide). This has been corrected in Version 2.0.

Unconverted DXF Objects: In TNM prior to Version 1.1, user-selected DXF objects had to be converted prior to saving and closing a run. Any unconverted DXF objects would then be deleted without recourse. Since Version 1.1, a Cancel option has been added so that you may cancel closing the run before the DXF objects are deleted.

2.1.5 Cleanup Run. A number of users had been experiencing "DB error" messages during computations. These DB (data base) errors were being caused by TNM's internal, third-party data base software (POET). Sometimes, simply acknowledging **OK** at the prompt when TNM encountered these errors would allow the model to continue computations uninterrupted. However, occasionally, the errors were too numerous for TNM to continue. As a result, a function has been implemented for the more *severe* cases. This function can be found as a new menu item called, Cleanup Run, in the File menu.

TNM User's Guide (Addendum)   *What's New*

To use Cleanup Run, select Cleanup Run in the File menu (see Figure 3). TNM will remind you to close your run and make a backup copy of it (see Figure 4). TNM will then display a **Cleanup Run** dialog which allows you to select the run (subdirectory) with the DB errors to "clean" (see Figure 5).

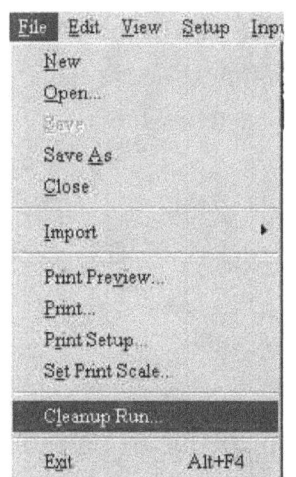

Figure 3. Cleanup Run menu item.

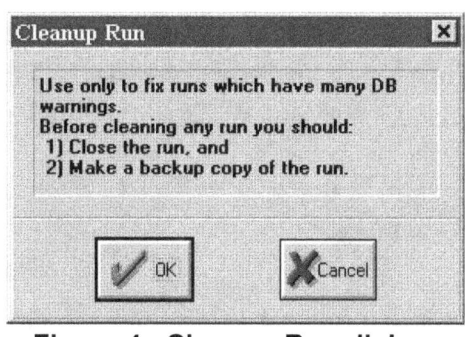

Figure 4. Cleanup Run dialog.

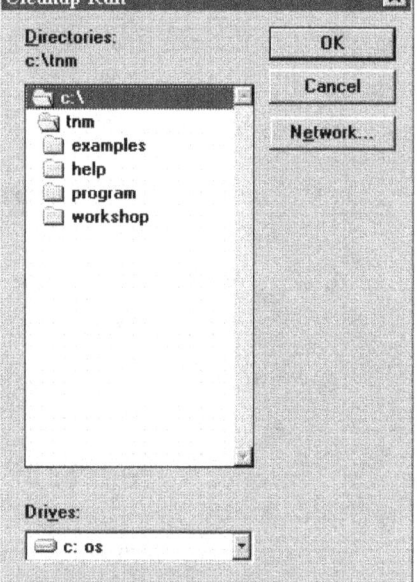

Figure 5. Cleanup Run "select run" dialog.

2.2 Edit Menu

2.2.1 Move. As mentioned in the TNM 1.0 User's Guide, the Move capability is not a menu option; it is only available in the Plan View using the Ctrl key and a mouse. This selection is used to graphically relocate a selected object, or portions of an object, to a new location, i.e., change the XY coordinates. Since Version 1.1, TNM has had added the ability for the user to also use the Snap tool in conjunction with Move. This aids the user in moving objects or points of objects and "snapping" them to other objects, e.g., placing a barrier on structure or a wall on top of a berm for a combination barrier.

What's New   TNM User's Guide (Addendum)

2.3 View Menu

2.3.1 Show/Hide. For all graphical views, any of TNM's input objects, as well as the aspects of those objects (e.g., point name and number), may be selected to be shown or hidden. This option is helpful in keeping complex cases from appearing cluttered and difficult to read. If point name/number text is not displayed after Show/Hide check boxes have been "checked," the TNM 1.0 User's Guide advises the user to first check their computer "TGLINK" settings. This was usually sufficient for most users - however, for some Windows NT Version 4.0 users, the following may also need to be checked:

1. Go to your computer's **Control Panel**.
2. Select and open the **System** icon.
3. When the **System Properties** dialog is displayed, select the **Environment** tab.
4. In the **Variable** field, type **TGLINK**.
5. In the **Value** field, type **C:\TNM\PROGRAM**
6. Click on the **Set** button and confirm that the settings that were just typed have been added to the **System Variables** and the **User Variables** window lists, respectively.
7. In the **System Variables** window, select **Path** in the **Variable** column. The path statement will then be highlighted and appear in the **Variable** and **Value** fields.
8. Click anywhere within the **Value** field. At the end of the text line, add **;C:\TNM**.
9. Click on the **Set** button. Click on the **OK** button to exit **System Properties**. Exit the **Control Panel**.

2.4 Setup Menu

2.4.1 General. General Input includes user preferences that affect TNM calculations. Most changes to TNM input will invalidate computed sound level results. However, in TNM prior to Version 1.1, General Input variables did not affect the sound level results. Since TNM Version 1.1, this has been modified to also invalidate results if the default ground type has been changed. **Use File, Save As to rename a run prior to making any changes, if you want to keep the original run with its already computed sound level results.**

> **Invalidating TNM results:** If you inadvertently changed the default ground type, thus invalidating previously computed results, you may close the run without saving changes. You may then reopen your run with your previously computed results restored.

2.5 Input Menu

 Input data changes: For all input dialogs, changes to data in a dialog's spreadsheet area are now reflected in RED until the user selects the Apply button to apply the changes.

2.5.1 Roadway Input. In the computation of the community noise equivalent level (L_{den}), a 5-dB penalty is added to evening operations, which equates to a 3.16 weighting factor (TNM Version 1.0). For consistency with state law in California, the primary user of the L_{den} metric, the weighting factor was changed to 3.00 in TNM Version 1.0b. For information on how to select L_{den} as your desired traffic entry type, refer to Section 7.2 in the TNM 1.0 User's Guide. For information on how to enter L_{den} traffic for roadways, refer to Section 8.3.4 in the TNM 1.0 User's Guide.

2.5.2 Receiver Input. The receiver input dialog has been changed such that information for all receivers are displayed on a single spreadsheet page (see Figures 6 through 9). In TNM prior to Version 1.1, the receiver input dialog displayed information for a single receiver (i.e., one receiver at a time). This change allows for more efficient editing of receiver input information.

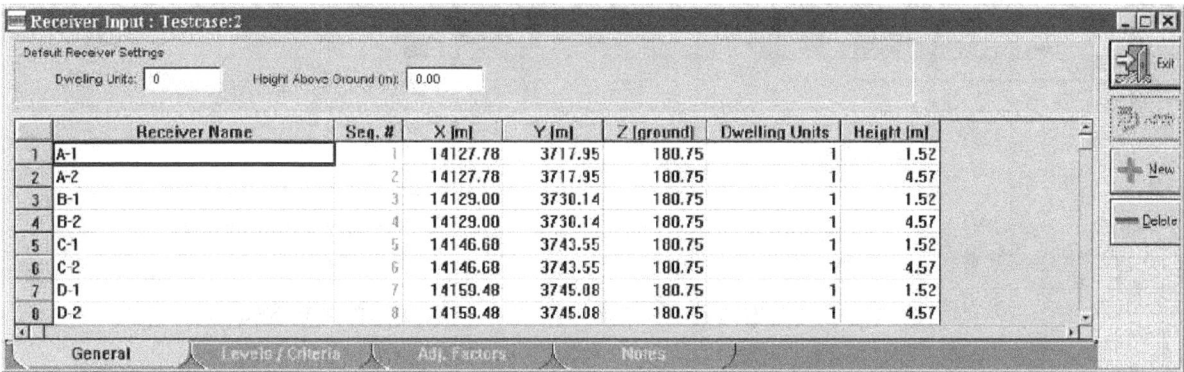

Figure 6. Changes to Receiver Input dialog: General tab.

 Receiver scrambling: In TNM Version 1.1, the new receiver dialog box was experiencing problems with runs which contained a large number of receivers resulting in some receiver information being scrambled. This has been corrected in Version 2.0.

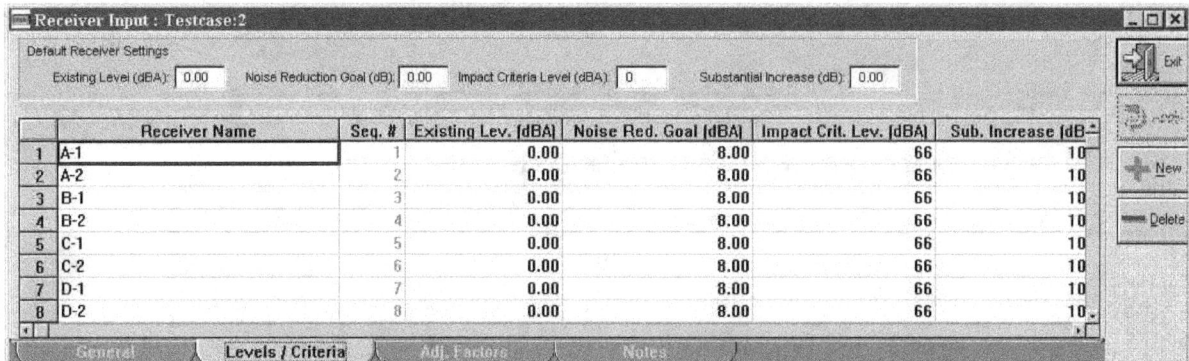

Figure 7. Changes to Receiver Input dialog: Levels/Criteria tab.

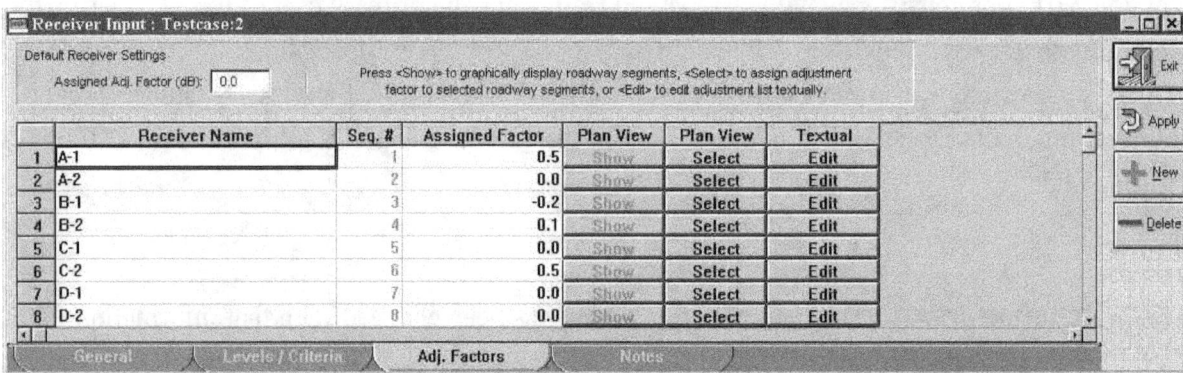

Figure 8. Changes to Receiver Input dialog: Adjustment Factors tab.

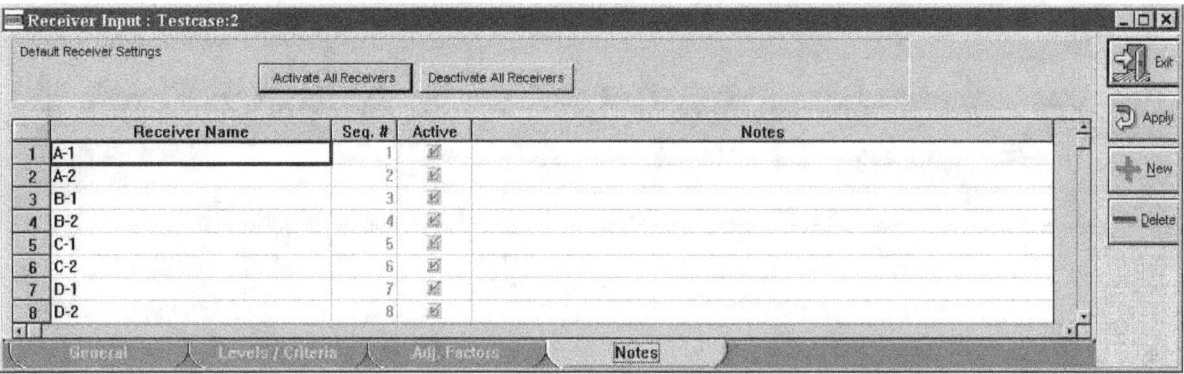

Figure 9. Changes to Receiver Input dialog: Notes tab.

In the Notes tab shown in Figure 9, a new attribute has been added - users are now able to **"activate"** or **"deactivate"** specific receivers for calculations by checking/unchecking boxes in the Active column. Two buttons are also available in the header area to activate and deactivate receivers. All receivers are active by default. Users can use this new feature to:

- Compute a newly added receiver(s);
- Compute a specific receiver(s) to see how a small input change affects that receiver(s); and

TNM User's Guide (Addendum) — What's New

- Compute a subset of receivers representing a portion of the study to save run-time.

Receiver active/inactive status is displayed in the receiver input table (see Figure 10) and is reflected in the sound level results tables, as well (see Figure 11).

Receiver Name	No.	#DUs	Coordinates (ground) X (m)	Y (m)	Z (m)	Height above Ground (m)	Existing LAeq1h (dBA)	Impact Criteria LAeq1h (dBA)	Sub'l (dB)	NR Goal (dB)	Active in Calc.
A-1	1	1	14,127.8	3,718.0	180.75	1.52	0.00	66	10.0	8.0	Y
A-2	2	1	14,127.8	3,718.0	180.75	4.57	0.00	66	10.0	8.0	Y
B-1	3	1	14,129.0	3,730.1	180.75	1.52	0.00	66	10.0	8.0	Y
B-2	4	1	14,129.0	3,730.1	180.75	4.57	0.00	66	10.0	8.0	
C-1	5	1	14,146.7	3,743.6	180.75	1.52	0.00	66	10.0	8.0	
C-2	6	1	14,146.7	3,743.6	180.75	4.57	0.00	66	10.0	8.0	
D-1	7	1	14,159.5	3,745.1	180.75	1.52	0.00	66	10.0	8.0	Y
D-2	8	1	14,159.5	3,745.1	180.75	4.57	0.00	66	10.0	8.0	Y

Figure 10. Changes to receiver input table.

Volpe Center — Acoustics Facility — 21 March 2002 — TNM 2.0
RESULTS: SOUND LEVELS
PROJECT/CONTRACT:
RUN: FHWA TNM Test Case
BARRIER DESIGN: INPUT HEIGHTS
ATMOSPHERICS: 20 deg C, 50% RH

Average pavement type shall be used unl a State highway agency substantiates the of a different type with approval of FHWA.

Receiver Name	No.	#DUs	Existing LAeq1h Calculated (dBA)	No Barrier LAeq1h Calculated (dBA)	Crit'n (dBA)	Increase over existing Calculated (dB)	Crit'n Sub'l Inc (dB)	Type Impact	With Barrier Calculated LAeq1h (dBA)	Noise Reduction Calculated (dB)	Goal (dB)
A-1	1	1	0.0	68.6	66	68.6	10	Snd Lvl	58.2	10.4	
A-2	2	1	0.0	73.0	66	73.0	10	Snd Lvl	62.4	10.6	
B-1	3	1	0.0	71.0	66	71.0	10	Snd Lvl	59.8	11.2	
B-2	4	1	0.0	0.0	66	0.0	10	inactive	0.0	0.0	
C-1	5	1	0.0	0.0	66	0.0	10	inactive	0.0	0.0	
C-2	6	1	0.0	0.0	66	0.0	10	inactive	0.0	0.0	
D-1	7	1	0.0	74.4	66	74.4	10	Snd Lvl	63.4	11.0	
D-2	8	1	0.0	74.6	66	74.6	10	Snd Lvl	63.9	10.7	

Figure 11. Changes to sound level results table.

2.5.3 Contour Zone. In TNM prior to Version 1.1, changes to contour zone input invalidated computed sound level results. Changes to contour zone coordinate input no longer invalidate results.

2.5.4 Adjustment Factors. Receiver adjustment factors are sound level adjustments in dB that are algebraically added to, not subtracted from, the sound levels calculated by TNM. In TNM prior to Version 1.1, TNM did not allow negative adjustment factors. Negative adjustment factors are now acceptable since Version 1.1. See Section 8.4.4 in the TNM 1.0 User's Guide for more information.

See Appendix A in the TNM 1.0 User's Guide for FHWA policy related to adjustment factors.

2.5.5 Input Check. The following new input checks have been implemented:
- Single point objects - TNM will display an input check error for objects that consist of only one point with the exception of receivers, and for zones that consist of only two points;
- Objects overlapping roadways - TNM will display an input check error for objects which overlap a roadway including the roadway's width; and
- Receivers with zero coordinates - TNM will display an input check error for receivers with all zero (x, y, and z) coordinates.

2.5.6 User-Defined Vehicles. The TNM User-Defined Vehicle's function (see Section 8.11 in the TNM 1.0 User's Guide) requests that the user enter four parameters as follows:
- *Similar TNM Type*: The Similar TNM Type represents the vehicle type from the standard list of five TNM vehicle types (i.e., autos, medium trucks, heavy trucks, buses, and motorcycles) which is most closely aligned with the user-defined vehicle. The assignment should be based on similarities in subsource heights, acceleration characteristics, and frequency spectrum;
- *Minimum Level*: The Minimum Level represents the emission level at idle and at very low speeds as defined by the engine/exhaust noise ("C" coefficient);
- *Slope*: The Slope ("A" coefficient) is determined from the analyses of the user's emission level measurements data; and

- ***Reference Level***: The Reference Level represents the emission level for the user-defined vehicle measured at 50 mph at a distance of 50 ft from the center of the near travel lane.

These parameters were intended to be used by TNM to compute an additional coefficient, the "B" coefficient (see Development of Reference Energy Mean Emission Levels for the FHWA Traffic Noise Model, Page 97, or Measurement of Highway-Related Noise, Page 75).[3,5] This "B" coefficient would then be used with the "A" and "C" coefficients to complete the TNM REMEL equation which computes the final source emission level.

However, TNM incorrectly assumes that the value entered for the Reference Level is synonymous with the "B" coefficient. Thus when the final source emission level is computed it is incorrect because the value it uses for the "B" coefficient is the user-defined Reference Level.

Although, it is intended that the "B"coefficient will be computed automatically by TNM, the interim solution is to enter the "B" coefficient determined from your preliminary analysis as the Reference Level. As before, enter the "C" coefficient as the Minimum Level, the "A" coefficient as the Slope. If a preliminary analysis was not performed, the "B" coefficient is easily found from the standard emission level equation since all other variables in the equation are known. Be sure to substitute the emission level at 50 mph into the equation when determining the "B" coefficient.

2.6 Calculate Menu

2.6.1 Error-Catching Mechanism.

An error-catching mechanism was first implemented in TNM Version 1.0a to eliminate any fatal crashes which users had been experiencing. Since then, additional error-catching mechanisms have been implemented.

As such, when TNM encounters an error during computations, the program skips the problematic receiver and continues computations with the next receiver. At the end of computations, TNM will inform the user of receivers that were invalidated either via a pop-up dialog (see Figure 12) when computing a single run or in the error report

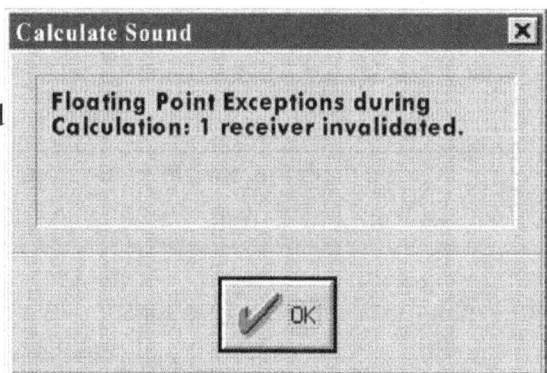

Figure 12. Invalidated receivers due to floating point errors.

generated when computing multiple runs (see Section 2.6.2). Any skipped receiver will be identified in TNM output tables as "invalid."

 Runs with errors: Any runs with receivers invalidated should be sent to the Volpe Center Acoustics Facility for further testing and diagnosis. A run consists of an OBJECTS.DAT and an OBJECTS.IDX file. Also provide an indication of which receiver the error occurred on and a detailed description of the error message.

Volpe Center Acoustics Facility Telephone: (617) 494-2372
55 Broadway, DTS-34 Email: Lau@Volpe.dot.gov
Cambridge, MA 02142

2.6.2 Multiple Runs. When the user selects Calculate, two menu options are now available (see Figure 13): calculate the sound levels for the Current Run; or calculate the sound levels in batch-mode for Multiple Runs.

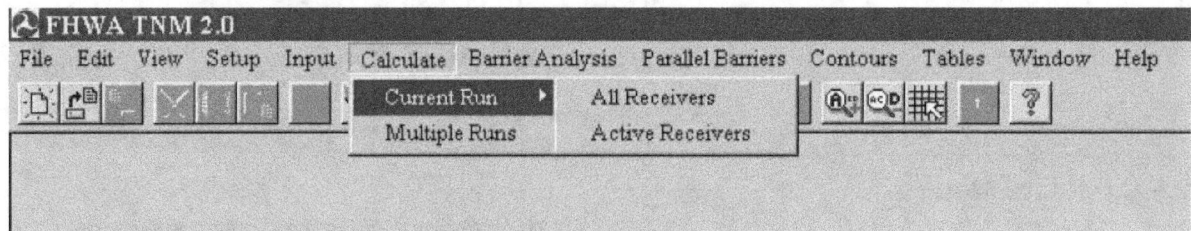

Figure 13. Calculate menu.

 When calculating multiple runs, you must first close all open runs or else TNM will display a warning.

- When the user selects Current Run, two options are available: All Receivers and Active Receivers. Selecting All Receivers begins TNM computations immediately on the currently open run as in previous versions of TNM prior to TNM Version 1.1. Selecting Active Receivers begins TNM computations on the currently open run for only the receivers marked active in the Receiver Input dialog (see Section 2.4.1).
- When the user selects Multiple Runs, the **Calculation Manager** dialog is displayed.

In the **Calculation Manager** dialog (see Figure 14), the *Run Name* area displays the list of selected runs. Runs are displayed with their full path name. The list will be automatically scrolled during the batch-mode calculations to ensure that the current run is visible in the list.

Buttons are provided on the right side of the **Calculation Manager** dialog for the following functions:

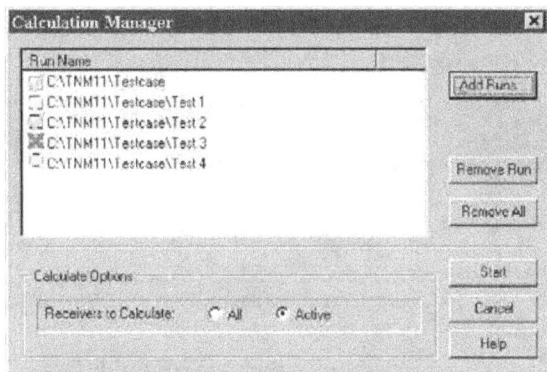

Figure 14. Calculation Manager dialog.

- *Add Run* - Displays the **Browse for Folder** dialog (see Figure 15) for adding runs to the list of runs to calculate;
- *Remove Run* - Removes a run selected in the **Run Name** list;
- *Remove All* - Clears the entire **Run Name** list;
- *Start/Continue* - Begins calculating runs. The button is greyed and un-selectable once calculations have started. If a calculation is cancelled, then this button will be changed to a *Continue* button, which when selected, will allow the interrupted calculation to continue (see Figure 16). If any changes are made to the **Calculation Manager** dialog, then this button will revert to a *Start* button to re-initiate calculations from the first run. Runs will be calculated in the order in which they were selected. As each run is calculated, a "gauge" dialog will be displayed to track the progress of that run (exactly as it does for a Current Run calculation);

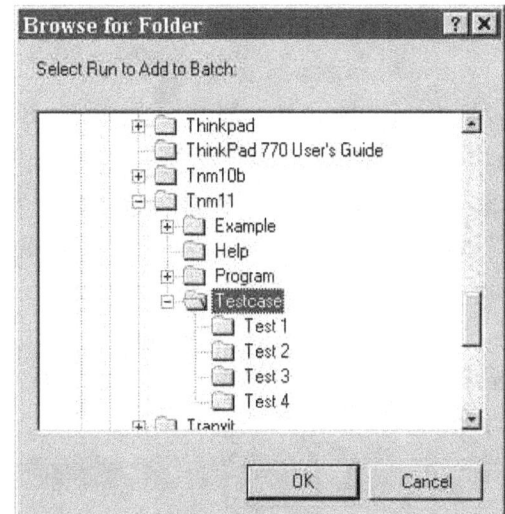

Figure 15. Browse for Folder dialog.

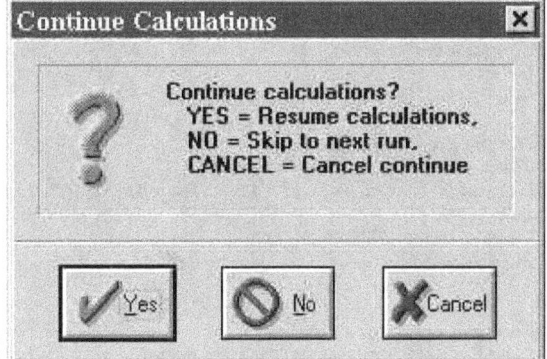

- *Cancel* - Exits the **Calculation Manager** dialog without calculating further runs but will retain the results of any runs already calculated; and

Figure 16. Continuing calculations after cancelling.

- *Help* - Displays help information regarding the **Calculation Manager** dialog.

A column of icons will be displayed to the left of the list of runs to indicate the status of each run as follows:

- Run not yet calculated;
- Computations in progress;
- Computations finished;
- Computations finished, but some receivers were invalidated (see Section 2.5.1);
- Computations cancelled; and
- Computations failed, errors were detected during input check that need fixing (see Appendix C of the TNM 1.0 User's Guide).

The status is also reported in an **Output Report** window which is generated during multiple-run calculations (see Figure 17). The window displays information from each run as the batch progresses, including the status of the run currently being calculated, the status of runs completed, and each run's total run-time. This report will also indicate if TNM encountered an error during computations, and, if so, how many receivers were invalidated for a particular run (see Section 2.6.1). TNM saves a copy of this report in a file called, batchCalc.out, and places it in your TNM/Program subdirectory. This file is over-written each time batch-mode calculations are performed.

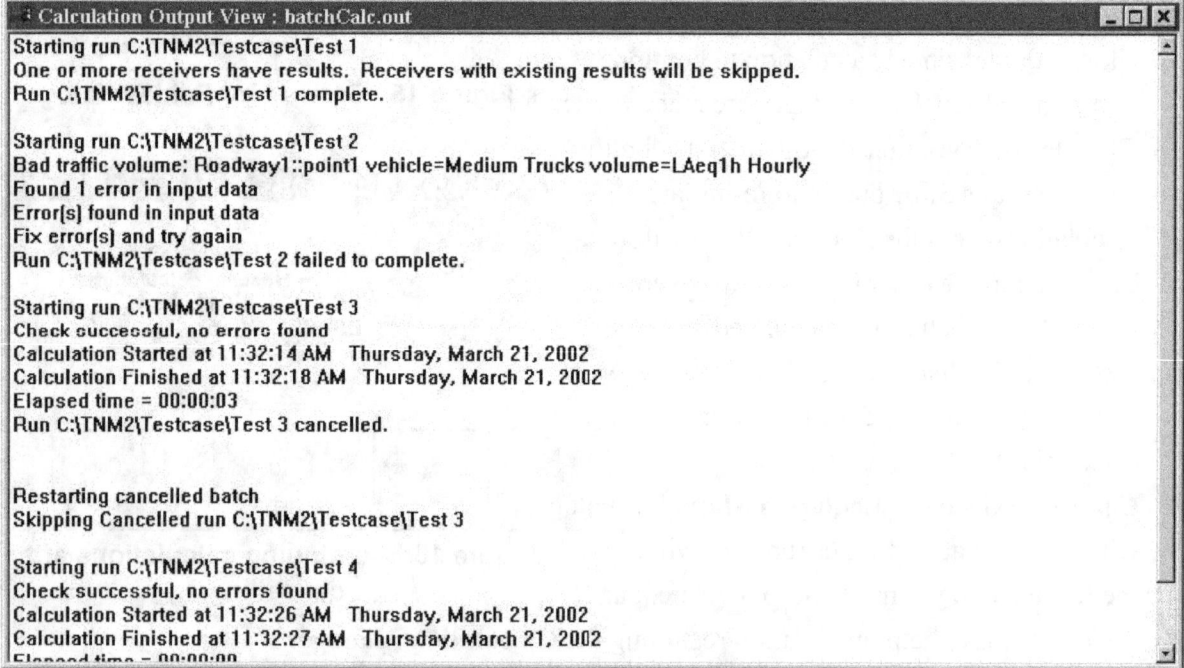

Figure 17. Multiple run (batch-mode) capability: output report.

2.7 Contours Menu

2.7.1 Calculating Contours. The previous version of NMPlot was having difficulties running on very fast computers (typically faster than 200 MHz) giving some users a runtime error message. The most recent MS-DOS version of NMPlot (Version 3.06) avoids these difficulties, and, as such, was updated with TNM Version 1.1.

Development is underway to implement the Windows version of NMPlot (Version 4.7) into TNM. In the interim, for users who would like to use some of the additional features that can be found in the Windows Version of NMPlot, it is available on the NMPlot website, http://www.wasmerconsulting.com/nmplot.htm. To use Version 4.7, download and run install_nmplot.exe. Once TNM has computed the grid file portion of NMPlot computation, that grid file may be viewed in Version 4.7 external to TNM.

"NMPlot Out of Memory" Error: If your computer displays an "Out of Memory" error while running NMPlot, modify your computer's memory allocation for MS-DOS applications as follows: open an MS-DOS window; select the MS-DOS window's upper-left icon; select Properties in the resultant pull-down menu; select the Memory tab; then increase the values shown. You may need to refer to your system administrator for additional assistance.

"Noisemap Grid File Could Not Be Opened" Error: NMPlot is DOS application with a 8-character filename/directory limit. If your TNM run is embedded within many subdirectories or has long subdirectory names, then NMPlot may display this error. Move your run out to the main C: directory to run contours.

2.8 Tables Menu

2.8.1 All Results Tables. In the right header area, TNM displays the current date and the version of TNM that is being used. An additional line of text has been added to let the user know which version of TNM was used to calculate the results being displayed (see Figure 18).

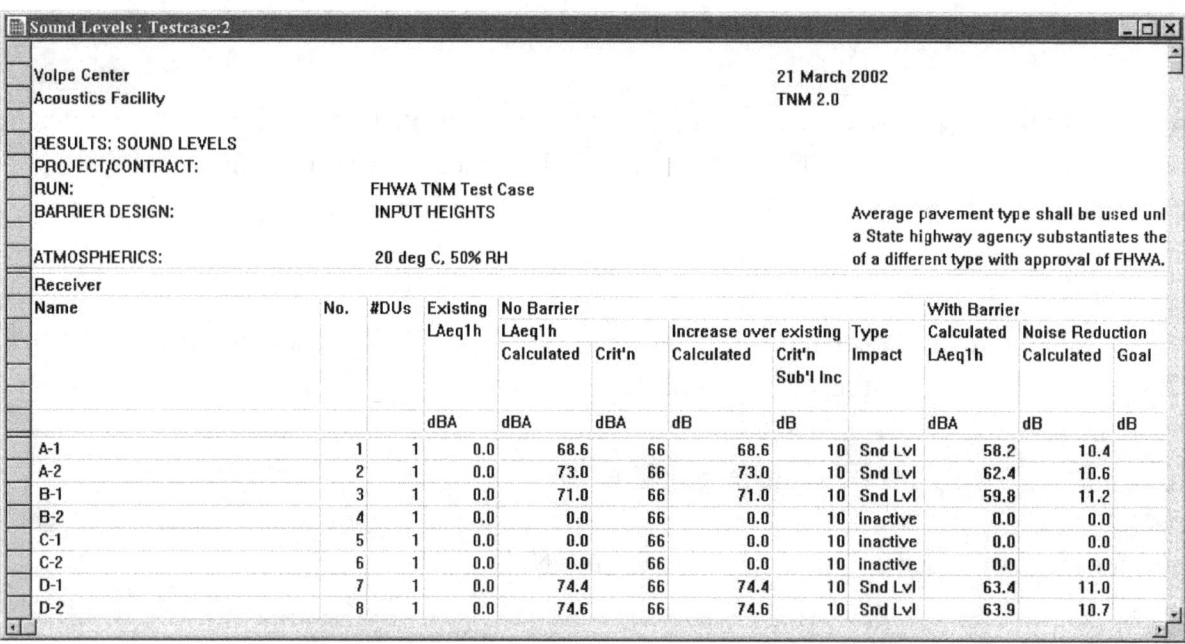

Figure 18. Version identification in all results tables.

2.8.2 Print Tables. A new menu item in the Tables menu, Print Tables, has been implemented to allow the user to print all TNM tables at once (see Figures 19 and 20). When the user selects this menu item, TNM displays a **Print Tables** dialog that lists all TNM tables (labeling each as an Input or Results table). The user can individually select tables by highlighting/clicking on the them in the list.

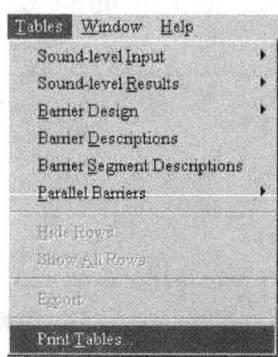

Figure 19. Print Tables menu item.

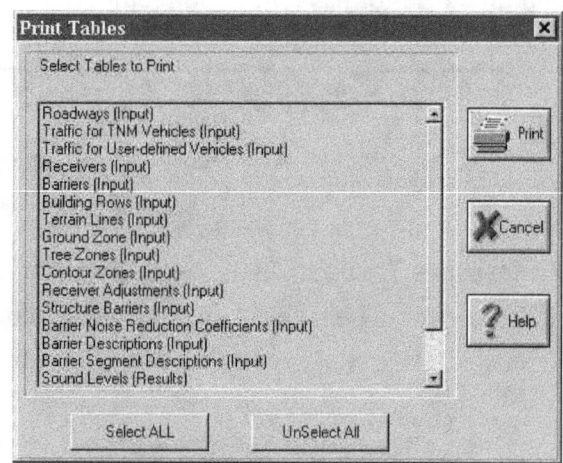

Figure 20. Print Tables dialog.

2.8.3 Barrier Design Table. A new table has been designed and incorporated into TNM. The Barrier Design Table was developed to aid the user during the barrier analysis and design process by placing the most frequently referred to information from various tables into one comprehensive table. The Barrier Design Table combines select information from the following tables:

- Sound Level Results;
- Barrier Descriptions;
- Barrier Segment Descriptions; and
- Diagnosis by Barrier Segment.

The table is available in two forms and can be selected using the Tables menu (see Figure 21). Two submenu options are listed: Barrier Design Table and Show Important Segments. Note that the Show Important Segments option is greyed-out until the user first opens a barrier design table. When the user opens a barrier design table, the table is displayed with the selected barrier analysis receivers and their associated data on sound levels, noise reduction, important barriers and barrier segments, and partial sound levels (see Figure 22). The user may then select the Show Important Segments option, which "unchecks" the option in the menu and toggles the table to a more condensed form, hiding the important barrier segment rows and their associated partial sound levels (see Figure 23).

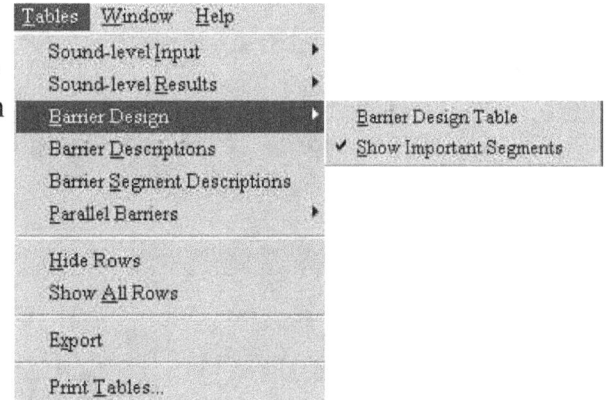

Figure 21. Barrier Design Table menu item.

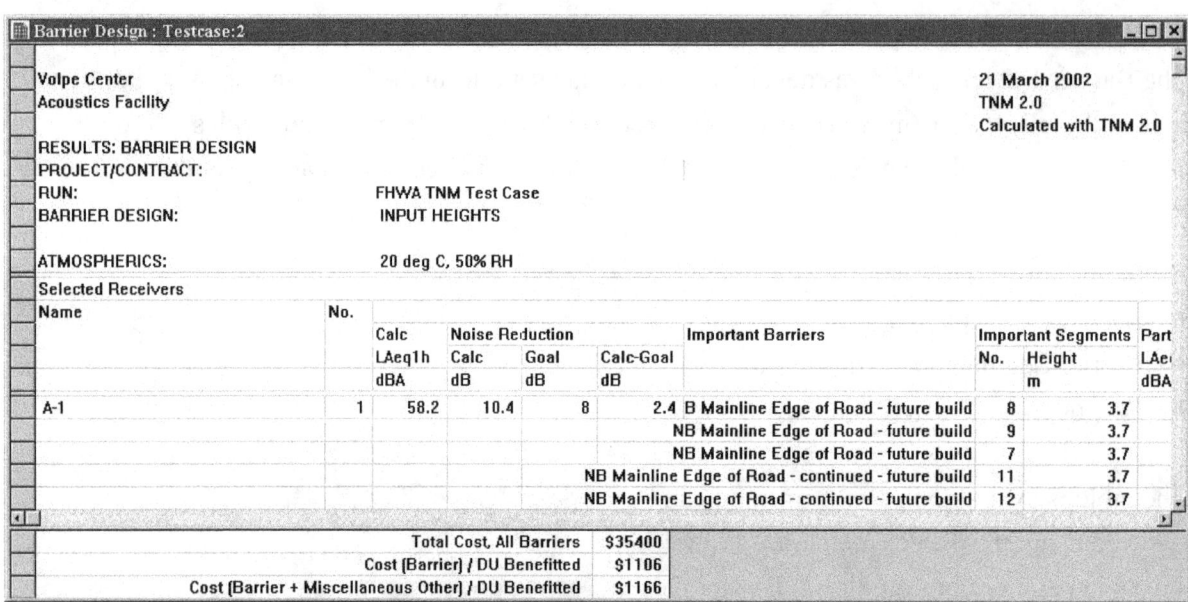

Figure 22. Barrier design table: expanded display.

Figure 23. Barrier design table: condensed display.

3. CERTIFIED OUTPUT FOR THE OFFICIAL TNM TEST CASE

This section contains the certified output computed for the official TNM test case using Version 2.0. Since TNM Version 1.0, updates to the model have resulted in an average 0.1-0.2 dB difference at receivers for some geometries. Appendix E in the TNM 1.0 User's Guide displays the sound level results computed using TNM Version 1.0 for the official TNM test case. The sound level results computed using the current version are shown in Figure 24 below.

Receiver Name	No.	#DUs	Existing LAeq1h dBA	No Barrier LAeq1h Calculated dBA	Crit'n dBA	Increase over existing Calculated dB	Crit'n Sub'l Inc dB	Type Impact	With Barrier Calculated LAeq1h dBA	Noise Reduction Calculated dB	Goal dB	Calculated minus Goal dB
A-1	1	1	0.0	68.6	66	68.6	10	Snd Lvl	58.2	10.4	8	2.4
A-2	2	1	0.0	71.0	66	71.0	10	Snd Lvl	60.4	10.6	8	2.6
B-1	3	1	0.0	71.0	66	71.0	10	Snd Lvl	59.8	11.2	8	3.2
B-2	4	1	0.0	72.7	66	72.7	10	Snd Lvl	62.3	10.4	8	2.4
C-1	5	1	0.0	72.8	66	72.8	10	Snd Lvl	61.4	11.4	8	3.4
C-2	6	1	0.0	74.7	66	74.7	10	Snd Lvl	64.2	10.5	8	2.5
D-1	7	1	0.0	72.3	66	72.3	10	Snd Lvl	61.3	11.0	8	3.0
D-2	8	1	0.0	74.6	66	74.6	10	Snd Lvl	64.0	10.6	8	2.6
E-1	9	1	0.0	72.6	66	72.6	10	Snd Lvl	60.9	11.7	8	3.7
E-2	10	1	0.0	73.8	66	73.8	10	Snd Lvl	63.8	10.0	8	2.0
F-1	11	1	0.0	72.6	66	72.6	10	Snd Lvl	60.9	11.7	8	3.7
F-2	12	1	0.0	74.0	66	74.0	10	Snd Lvl	63.8	10.2	8	2.2
G-1	13	1	0.0	73.7	66	73.7	10	Snd Lvl	61.4	12.3	8	4.3
G-2	14	1	0.0	74.7	66	74.7	10	Snd Lvl	64.7	10.0	8	2.0
19aE-1	15	1	0.0	74.4	66	74.4	10	Snd Lvl	61.7	12.7	8	4.7
H-1	16	1	0.0	73.5	66	73.5	10	Snd Lvl	61.3	12.2	8	4.2
H-2	17	1	0.0	74.5	66	74.5	10	Snd Lvl	64.7	9.8	8	1.8
I-1	18	1	0.0	73.4	66	73.4	10	Snd Lvl	61.1	12.3	8	4.3
I-2	19	1	0.0	74.2	66	74.2	10	Snd Lvl	64.5	9.7	8	1.7
J-1	20	1	0.0	72.1	66	72.1	10	Snd Lvl	60.4	11.7	8	3.7
J-2	21	1	0.0	73.1	66	73.1	10	Snd Lvl	63.3	9.8	8	1.8
K-1	22	1	0.0	71.5	66	71.5	10	Snd Lvl	60.1	11.4	8	3.4
K-2	23	1	0.0	72.7	66	72.7	10	Snd Lvl	62.9	9.8	8	1.8
L-1	24	1	0.0	71.2	66	71.2	10	Snd Lvl	59.9	11.3	8	3.3
L-2	25	1	0.0	72.6	66	72.6	10	Snd Lvl	62.8	9.8	8	1.8
M-1	26	1	0.0	70.7	66	70.7	10	Snd Lvl	59.6	11.1	8	3.1
M-2	27	1	0.0	72.3	66	72.3	10	Snd Lvl	62.4	9.9	8	1.9
19aE-2	28	1	0.0	73.8	66	73.8	10	Snd Lvl	61.1	12.7	8	4.7
N-1	29	1	0.0	69.3	66	69.3	10	Snd Lvl	59.1	10.2	8	2.2
N-2	30	1	0.0	71.3	66	71.3	10	Snd Lvl	61.8	9.5	8	1.5
O-1	31	1	0.0	72.2	66	72.2	10	Snd Lvl	60.5	11.7	8	3.7
O-2	32	1	0.0	74.2	66	74.2	10	Snd Lvl	64.1	10.1	8	2.1

Dwelling Units	# DUs	Noise Reduction Min dB	Avg dB	Max dB
All Selected	32	9.5	10.9	12.7
All Impacted	32	9.5	10.9	12.7
All that meet NR Goal	32	9.5	10.9	12.7

Figure 24. Updated sound level results table for the official TNM test case.

REFERENCES

1. Anderson, Grant S., Cynthia S.Y. Lee, Gregg G. Fleming. <u>FHWA Traffic Noise Model,® Version 1.0: User's Guide</u>. Report No. FHWA-PD-96-009 and DOT-VNTSC-FHWA-98-1. Cambridge, MA: John A. Volpe National Transportation Systems Center, Acoustics Facility, January 1998.

2. Menge, Christopher W., Christopher J. Rossano, Grant S. Anderson, Christopher J. Bajdek. <u>FHWA Traffic Noise Model,® Version 1.0: Technical Manual</u>. Report No. FHWA-PD-96-010 and DOT-VNTSC-FHWA-98-2. Cambridge, MA: John A. Volpe National Transportation Systems Center, Acoustics Facility, February 1998.

3. Fleming, Gregg G., Amanda S. Rapoza, Cynthia S.Y. Lee. <u>Development of National Reference Energy Mean Emission Levels for the FHWA Traffic Noise Model,® Version 1.0</u>. Report No. FHWA-PD-96-008 and DOT-VNTSC-FHWA-96-2. Cambridge, MA: John A. Volpe National Transportation Systems Center, Acoustics Facility, November 1995.

4. Bowlby, William, Theodore Patrick, Cynthia S.Y. Lee, Gregg G. Fleming. <u>FHWA Traffic Noise Model,® Version 1.0: Trainer CD-ROM</u>. Cambridge, MA: John A. Volpe National Transportation Systems Center, Acoustics Facility, March 1998.

5. Lee, Cynthia S.Y., Gregg G. Fleming. <u>Measurement of Highway-Related Noise</u>. Report No. FHWA-PD-96-046 and DOT-VNTSC-FHWA-96-5. Cambridge, MA: John A. Volpe National Transportation Systems Center, Acoustics Facility, May 1996.

www.ingramcontent.com/pod-product-compliance
Lightning Source LLC
Chambersburg PA
CBHW081808170526
45167CB00008B/3381